The Obesity Remedy

Unlocking Secrets to lasting weight loss, conquering food cravings, and energizing your life.

By

MARY D. MILLER

DISCLAIMER

TABLE OF CONTENTS

CHAPTER FOUR: APPROACHES TO PREVENT OBESITY

CHAPTER FIVE: HOW EXCESSIVE SUGAR CONSUMPTION CAN LEAD TO OBESITY AND FAT AMOUNT, SPECIFICALLY IN THE ABDOMINAL AREA

INTRODUCTION

DEFINITION OF OBESITY

Obesity is a disease characterized by excessive body fat that raises the chance of developing health issues. A common cause of Obesity is consuming more calories than you burn via exercise and everyday activities. Obesity is an abnormal or excessive accumulation of fat that puts your health in danger.

When you have a body mass index that reaches 25, it is considered overweight. Whereas Obesity is identified by a BMI that exceeds 30. In accordance with the global burden of disease, the problem has reached epidemic proportions, causing over 4 million annual deaths in 2017 due to excess weight and Obesity.

A complex illness characterized by having too much body fat is obesity. Obesity is a medical condition more than simply a visual issue that increases the risk of many illnesses and other health problems. Among these can be diseases, high cholesterol, heart disease, sleep apnea, high blood pressure, liver disease and certain types of cancer.

The Obesity Remedy

The encouraging news is that even a minor weight loss can assist in the health problems related with Obesity. Weight loss can be achieved through healthy eating, improved physical activity, and cultivating appealing habits.

Alternatives for controlling Obesity include prescription medications and weight loss techniques. connected with having too much body fat, the obesity epidemic in the United States is a serious public health concern that could result in high medical expenses.

While people are aware of the significant risk of diabetes mellitus and cardiovascular disease brought on by having too much body fat, there are a variety of additional health issues that can coexist with being overweight or obese, potentially bringing on early morbidity and mortality. The public has to see obesity as a serious health concern and not just a matter of appearance or lifestyle choice.

More frequent obesity screenings and discussions about healthy weight management will be made possible by a greater understanding of the health risks associated with being overweight.

This could lead to a greater commitment of healthcare resources to efficient obesity prevention and management strategies.

CHAPTER ONE: OBESITY CAUSES AND FACTORS

Genetic influence

Genetics may contribute to obesity, according to evidence, but genes don't explain everything. People with no family history or genetic predisposition to being overweight can battle with weight loss for the rest of their lives, whereas people with such traits can live their entire lives at a healthy weight.

The genes are linked to three main subtypes of obesity: polygenic, syndromic, and monogenic. Severe class 3 obesity known as "monogenic obesity" develops without any developmental abnormalities. Obesity that coexists with developmental disorders like Down syndrome is referred to as syndromic obesity.

Everyone is susceptible to polygenic obesity, sometimes referred to as common obesity. There are now roughly 20 genetic alterations known to cause monogenic obesity. Genetics is also involved in syndromic obesity.

For instance, a chromosomal defect results in Prader-Willi syndrome (PWS), a hereditary syndrome. Because of central nervous system malfunction that causes an abnormally high hunger (hyperphagia), people with this disease have a propensity to become obese at a young age.

The likelihood of genetic defects causing obesity is extremely low, despite the possibility. Your predisposition to acquire weight and develop obesity is largely determined by environmental circumstances rather than genetics. Obesity does, however, frequently run in families. The environment is frequently to blame for this. People who share housing and meals frequently develop the same lifestyles and eating habits.

The part genes play in determining metabolism

You probably know someone who can eat everything they want without gaining weight. Although it is impossible to consume an infinite number of calories without gaining weight, everyone has a different metabolic rate, and the higher your metabolic rate, the more you can consume without packing on the pounds. The inverse is also accurate. It could seem like all it takes to gain weight if your metabolic rate is lower than average. Although this statement is obviously exaggerated, it is true that two individuals with the same height, weight, body type, age, and gender can burn calories at different rates.

Your metabolic rate could be influenced by your genes. This is identified as your basal metabolic rate, also referred to as BMR. The quantity of energy (calories) your body burns while you're at rest is known as your basal metabolic rate, or BMR. Your genes play a role in determining some aspects of this rate, but other factors can also affect BMR.
One of those factors is muscle mass. Your BMR will increase as your body's percentage of muscle increases.

A male who is 5'7" and weighs 160 pounds will likely have a higher BMR than a woman who is the same height and weight because men tend to have more muscle mass. Although it is a generalization, it serves to show how muscle mass affects BMR.
Ultimately, your environment and lifestyle choices will have the biggest impact on your weight (unless

you have uncommon genetic or health issues). Even those with slower metabolisms can keep their weight under control, and those with faster metabolisms can become overweight. So, even if BMR has a hereditary component, environment and lifestyle are better predictors of obesity than genetics.

one's thyroid

Your metabolism is significantly influenced by your thyroid gland. Additionally, certain inherited thyroid conditions can cause your BMR to increase or decrease. Your metabolism will be slower if your thyroid gland is underactive (hypothyroidism), and faster if it is overactive (hyperthyroidism).

If you have hypothyroidism, you should work with your physician to raise the levels of thyroid hormones in your blood. Low thyroid hormone levels can prevent you from losing weight and have a number of unsettling and even harmful side effects. Thyroid hormone replacement is the most typical treatment for an underactive thyroid gland. If you have thyroid hormones on prescription, take them as prescribed each day.

Your metabolism should recover after your hormone levels return to normal.

The part that genes play in determining appetite

Some hereditary syndromes, as we already discussed, tend to enhance appetite, among other things. One of those syndromes is Prader-Willi syndrome. People with PWS have a central nervous system disorder that makes them more ravenous. This uncommon illness demonstrates the possibility of a genetic-appetite relationship.

There is more proof that certain genes and hunger are related. According to a UCLA College of Medicine study, some persons who are chemically dependent on alcohol have a genetic signature with people who tend to binge on carbohydrates.

The potential to have taste receptors that are sensitive to sweetness was shown to be correlated with a genetic mutation, according to a study on sweet sensitivity. There is evidence from other research that obesity runs in families, but it is unclear if this association is caused by genes, environment, or both. Although there is some evidence that your genes may influence your hunger or appetite, there is more support for the idea that lifestyle has a greater impact than genetics. In actuality, your DNA is constantly evolving.

Furthermore, some elements of your DNA code are modifiable. It may change in response to your surroundings, your feelings, what you eat and drink, and other factors. This theory's underlying field of study is known as epigenetics. According to epigenetics experts, your lifestyle decisions and mentality can change the epigenetic code in your body. There is proof that your behavior can influence how your genes are expressed. Research is being done to see whether nutrition may be used to turn genes on or off, affecting how much your genes affect your life.

This is a very simplistic exposition of a difficult subject, and the science of epigenetics is still in its infancy. But the truth is that while your genes may affect your weight, your lifestyle, and environment are generally much more significant determinants. You may lower your risk of obesity and the illnesses linked with obesity by making healthy decisions, such as eating a balanced diet rich in nutrients and exercising frequently, and you can also overcome any inherited propensity for being overweight.

Obesity epigenetic causes

The term "epigenetics" refers to inheritable and reversible processes, such as DNA methylation, covalent histone modifications, chromatin folding, and, more recently, the regulation of miRNAs and polycomb group complexes, that alter gene expression without changing the underlying base pair sequence. The study of 'genome-wide' epigenetic changes is known as epigenomics. Epigenetics is widely applicable to many areas of biological research since variations in gene expression are essential for both healthy development and the progression of disease.

Role of environmental epigenetics

The health issue of obesity is on the rise. In recent years, it has been hypothesized that environmental factors, in addition to hereditary ones, play a major role in the vulnerability to obesity. Numerous studies have been done on the effects of environmental and lifestyle factors, including nutrition and bioactive dietary ingredients, on epigenetic phenomena like DNA methylation and different kinds of histone modifications. Epigenetic mechanisms play a significant role in the onset and intergenerational transmission of chronic non-communicable diseases (NCD), including obesity, because an individual's epigenetic patterns are established during early gestation and are subject to transformation by environmental factors throughout their lifetime.

Genomic imprinting in humans regulates more than 100 protein-coding genes. These unusual genes are arranged in chromosomal domains, each of which is governed by an "imprinting control region" (ICR) with variable levels of methylation. We are now beginning to understand how ICRs regulate the parental allele-specific expression of nearby genes.

This epigenetic method involves the activity of long non-coding RNAs at several imprinted domains. The fact that hundreds of microRNA and short nucleolar RNA genes are also expressed by imprinted gene domains and that these are the densest collections of small RNA genes in mammalian genomes is less widely known. It has been demonstrated that imprinted short RNAs regulate particular processes in metabolism and development, and that altering their expression results in disease.

Recent research suggests that in addition to requiring a healthy balance between energy intake and expenditure, normal metabolic regulation during adulthood can also be influenced by pre- and postnatal settings.

In reality, by affecting the epigenetic control of particular genes, maternal dietary restriction during pregnancy can change the metabolic phenotype of the offspring, which can then be passed on to the next generations. The methylation and/or histone acetylation levels of genes engaged in particular but also in more general metabolic processes have been reported to be altered in studies on the epigenetic marks in obesity.

The etiology of obesity is complicated and involves intricate interactions between the neuroendocrine condition, prenatal programming, genetic make-up, and other unfavorable environmental variables, like sedentarism or poor eating practices. Despite the relatively high heritability of common, nonsyndromic obesity (40–70%), epigenetic processes cannot entirely account for the heredity of obesity due to genetic susceptibility variations. Extreme forms of obesity, such as Prader-Willi syndrome (PWS), are known to result from perturbation in these systems, but vulnerability to obesity has also been conclusively linked to these mechanisms.

Additionally, environmental exposures throughout crucial developmental phases can alter the epigenetic mark profile and cause obesity.
Epigenetics has become a key factor in determining obesity among the several methods that it can be caused by, as was discussed above. It may be possible to forecast a person's propensity to gain or lose weight by identifying those people who exhibit changes in their DNA methylation profiles, particular histone modifications, or other epigenetically related processes.

The avoidance of excessive fat deposition, the identification of the best weight loss strategy, and the use of newer therapeutic strategies for obesity may all benefit from studies on the epigenetic pathways controlling weight homeostasis.

This emphasizes the need for a renewed investigation into the epigenetic causes of the obesity epidemic, as knowledge of the mechanisms enabling intergenerational metabolic reprogramming has significant ramifications for our comprehension of phenotypic diversity and evolution.

There is currently a lot of evidence indicating nutritional epigenetic programming of health and the response to diet itself, as well as a mechanism via which diet and nutrition can directly alter the DNA. It has been demonstrated that a variety of nutrients and food substances that are altered in the human diet result in stable epigenetic changes over a range of durations.

Understanding the significance of epigenetic mechanisms in the control of gene expressions as a result of gene-environment interactions has advanced significantly over the past 20 years. The transgenerational inheritance of epigenetic features makes nutrition, among many other environmental factors, a crucial player in the induction of epigenetic modifications not only in the directly exposed organisms but also in succeeding generations.

By directly blocking the enzymes that catalyze DNA methylation or histone modifications or by changing the availability of the substrates required for those enzymatic processes, nutrients and bioactive food ingredients can affect epigenetic events. Nutritional epigenetics have been seen in this context as a desirable method to avoid obesity and associated NCD.

Although the prospect of discovering a cure or preventing these diseases is exciting, nutritional epigenetics knowledge is currently limited. More research is required to increase the knowledge base and better understand how nutrients or bioactive food components can be used to maintain health and fend off diseases through modifiable epigenetic mechanisms.

An individual's epigenetic state changes from time to time, and there is a growing understanding of the significance of this variation in health and disease. The role of epigenetic control in how the genome responds to and engages with its environment, as well as how it may be able to change its environment via impacting behavior, is crucial. There is substantial evidence that eating has an impact on the epigenome. Acetyl and methyl groups are the substrates for epigenetic processes. The hope for nutritional epigenetics is that it will make it clearer how diet can affect health by having a direct impact on the DNA.

The process of epigenetics is dynamic. Clinical endpoints connected to epigenetic phenomena may have reversible gene expression as a mechanistic explanation. Metabolic programming via epigenetics has an impact on several significant health endpoints. These relate to several future health concerns of NCDs, including obesity, and involve the relationship between perinatal nutrition, nutritional epigenetics, and programming at an early developmental stage.

Thus, through diet-mediated epigenetic intervention at particular developmental periods, programming and eventually reprogramming can both show to be useful strategies for enhancing health. Because obesity in childhood has been linked to newborn weight and weight rise from the first months of life, epigenetic factors influencing weight gain soon after birth may be a good target for preventing obesity.

Long-term consequences

The detrimental effects on children of diabetic and/or obese mothers are passed on to the following generation in large part due to epigenetics. In comparison to spermatozoa from offspring of non-diabetic mothers, the DNA methylation of H19 in spermatozoa from diabetic moms is considerably higher. These findings suggested that obesity and/or pre-gestational diabetes can change DNA methylation in spermatozoa from children.

'Visceral obesity' appears to be developed by epigenetic processes. Specific metabolic changes connected to obesity are closely related to visceral fat formation. It has been investigated how DNA methylation affects outcomes related to weight loss, excess body weight, or adiposity. Significantly more information on the epigenomics of obesity and the distribution of body fat will become available in the near future as a result of the advancement of new sequencing and omics technologies.

Both the type and amount of diet can have an impact on epigenetic pathways. Dietary substances including the antioxidants sulforaphane from cruciferous vegetables and epigallocatechin-3-gallate from green tea have been shown to have an impact on the expression of noncoding RNA, histone alterations via histone deacetylase, and DNA methyltransferase. It has been demonstrated that controlling these epigenetic pathways can significantly affect how different NCDs develop and advance. It has been demonstrated that an "epigenetic diet" containing the aforementioned substances along with calorie restriction can affect cellular longevity by modifying a few important genes that encode telomerase and p16.

According to available data, the de novo DNA methyltransferase Dnmt3a functions in the neurons of the paraventricular nucleus of the hypothalamus to relate environmental factors to disturbed energy balance and the emergence of obesity.

Psychotropic medications, for example, are well known for their capacity to increase fat accumulation in psychiatric patients. Adipogenesis, the process of creating newly differentiated adipocytes, and the amount of lipid accumulation are the two factors that determine fat mass. Adipogenesis-related cell fate decisions and differentiation are heavily influenced by epigenetic programming. Parallel to this, other genetic models provide strong support for the effects of epigenetics on energy metabolism. As a result, a range of psychiatric drugs, frequently taken in combination and for extended durations, may make use of a similar epigenetic effector pathway, increasing adipogenesis or decreasing energy metabolism.

Particularly, the new finding that G protein-coupled signaling cascades can directly affect epigenetic regulatory enzymes suggests that psychiatric drug surface receptor function. The effects of the histone demethylase inhibitor and clinically approved antidepressant tranylcypromine also point to potential therapeutic applications, which, in the obese phenotype, have remarkable therapeutic benefits on metabolism.

Behavioral elements

Research on obesity has recently focused on behavioral factors or actions that may lead to weight increase through overeating or decreased physical activity, as well as environmental factors that may have an impact on health. A review of the literature identifies the particular eating habits linked to obesity. Frequent fast food intake, dining out frequently, big portion sizes, drinking a lot of sugary beverages, and skipping breakfast are all behaviors that contribute to obesity. Multistructural variables, such as the physical environment and socioeconomic situation, have been proven to have a considerable impact on food intake and energy expenditure in addition to these behavioral aspects.

The availability of recreational spaces for physical activity, the proximity of large supermarkets, the concentration of fast-food restaurants in any given area, and socioeconomic status are all examples of environmental influences that may favorably or unfavorably affect dietary behaviors and physical activity patterns. Large supermarkets, for instance, offer a wide range of healthy foods at affordable costs, which may have an impact on food shopping habits and the availability of healthier meals at home.

Last, but not least, there has been a lot of attention on promoting healthy eating in schools and the critical link between the family environment and nutrient intake.

External Variables

Environmental factors and a number of genes all contribute significantly to the development of obesity, which is a complex disease. Obesity has been linked to neurodevelopmental disorders including autism, schizophrenia, and fragile X syndrome as well as neurodegenerative illnesses like Alzheimer's, Parkinson's, and Huntington's diseases.

Physical activity, alcohol use, socioeconomic position, parental feeding practices, and food are a few environmental factors that contribute to obesity. It's interesting to see the similarities between some of these environmental factors and neurodegenerative and neurodevelopmental illnesses. Obesity hinders the neurodevelopment of skills including memory and fine motor control. Additionally, maternal obesity has an impact on the offspring's mental and cognitive development. Insulin resistance, pro-inflammatory cytokines, and oxidative damage are among the common molecular pathways linked to obesity and neurodegenerative/neurodevelopmental illnesses that can hinder brain development or cause cell death.

There are other elements besides obesogenic environmental circumstances that affect neurodegenerative and neurodevelopmental illnesses. In fact, a number of genes (LEP, LEPR, POMC, BDNF, MC4R, PCSK1, SIM1, BDNF, TrkB, etc.) implicated in the leptin-melanocortin pathway are linked to obesity as well as neurodegenerative and neurodevelopmental illnesses.

Additionally, in recent years, new genes linked to neurodegenerative or neurodevelopmental diseases, such as obesity (FTO, NRXN3, NPC1, NEGR1, MTCH2, and GNPDA2, among others), as well as neurodegenerative or neurodevelopmental diseases (APOE, CD38, SIRT1, TNF, PAI-1, TREM2, SYT4, FMR1, and TET3), have been discovered. In conclusion, a greater knowledge of the etiology of these disorders would result from research into the genes, obesogenic environmental factors, and gene-environment interactions.

Obesogenic Setting

Dogs exposed to an obesogenic environment (owned by obese individuals) demonstrated a higher prevalence of obesity compared to dogs with lean owners, demonstrating that an obesogenic environment may play a significant role in the development of obesity. The increase in this disease has been attributed, in part, to obesogenic environmental factors.

Risk factors for the development of common obesity include the environment (nurture) and the genes (nature). The connection between genes and environment can enhance the vulnerability to developing the disease, even though these two factors are typically investigated separately. The nucleotides that make up human DNA (deoxyribonucleic acid), also known as genes, are considered to be a component of nature.

Individual differences in a trait and genetic variation are caused by base pair changes in DNA. Environment, a non-genetic factor that may alter a trait, is a component of nurture.

Physical exercise

We must consider the two primary components of the energy balance—energy intake and expenditure—when discussing an obesogenic environment. A sedentary lifestyle, or physical inactivity brought on by long stretches of TV watching, combines with the genetic tendency that leads to the development of obesity in terms of lower energy expenditure.

In fact, drinking alcohol might lessen the impact of obesity genetic variations by lowering BMI. This is in line with studies that show alcoholic individuals have decreased levels of physical activity and BMI as a result of increased lipolysis and lipid metabolism abnormalities. In fact, it has been shown that prolonged alcohol use can cause lipodystrophy in rats, which is brought on by a disruption in lipogenesis. Increased lipolysis and fatty acid release will cause the fatty acids to be transported to the liver, where they will accumulate and eventually cause hepatic steatosis.

Status Socioeconomic

Numerous research have examined the link between obesity and socioeconomic position. According to reports, the prevalence of obesity rises when deprivation levels rise because of a poorer diet and less physical exercise. It's interesting to note that in wealthy nations, obesity prevalence was greater among the poor and remained stable among the richest.

However, in developing nations, the upper class has a higher prevalence of obesity and overweight. The emerging middle class—poor people who have become wealthier—has the highest risk of becoming fat in developing nations like Mexico.

Additionally, other "environmental layers"—including the intrauterine environment, mother-child interactions, the food and community environment, parental feeding habits, and others—could influence someone's propensity to become obese. These "environmental layers" indicate the biological or social factors that might cause anyone to become ill. We shall briefly discuss the feeding habits of parents and food (diet) as a component of energy intake, a crucial component of the energy balance, among these layers.

disease We shall briefly discuss the feeding habits of parents and food (diet) as a component of energy intake, a crucial component of the energy balance, among these layers.

Parental Feeding Practices

Parental feeding habits are one of the environmental factors influencing a child's appetite. Giving food as a reward for excellent deeds, for instance, is linked to a higher intake of harmful foods and drinks.

Children who have trouble controlling their eating have been linked to having heavier bodies. Preschoolers who are picky eaters frequently experience family tension and worry from their parents.

The relationship between authoritative parenting and their toddler's non-picky eating is favorable, indicating that this parent-feeding behavior may be able to solve the feeding issues. Mothers who encourage a favorable body image in their children are expected to have less weight; nevertheless, forcing children to eat has been linked to weight growth. The influence of nurture also affects how genes interact with the environment to influence a child's appetite and eating habits. Support a healthy body image in children, yet forcing children to eat has been linked to weight growth. The influence of nurture also affects how genes interact with the environment to influence a child's appetite and eating habits.

Diet

One of the key environmental elements that affects the development of obesity is diet. Dietary patterns have altered as a result of a faster lifestyle, leading people to purchase prepared foods rather than cooking their own meals. Additionally, a meta-analysis revealed that the consumption of soft drinks has increased in recent years and that added sugar was present in 75% of foods and beverages.

The risk of developing obesity, diabetes, and metabolic syndrome is therefore increased by consuming sugar-sweetened beverages. The favorable relationship between the consumption of soft drinks and fried foods and adiposity features is also linked to a hereditary susceptibility to obesity.

According to a review, consuming fried meals four or more times a week increases your chance of being obese and having other chronic diseases like type 2 diabetes (T2D) and hypertension, which can cause coronary artery disease.

Another study found that enhancing cardiometabolic profiles by increasing the intake of fruits and vegetables and reducing the intake of red and processed meats.

A larger intake of fruits, vegetables, legumes, nuts, and whole grains is advised, with less than 10% of total calories coming from free sugars and 30% coming from fats, according to WHO's published dietary recommendations. Despite the fact that these suggestions have been made public, it has been established that a number of factors, including household income and eating and cooking habits, make it difficult to consume healthy meals.

A recent study showed that diverse populations' diets are unsustainable, unequal, and unhealthy. People are eating "unhealthy," as previously mentioned, with a high intake of sugar and fats. A relationship between people and the environment is the food. Plant and animal production for human use is causing biodiversity loss, altering the environment, and is otherwise "unsustainable." Finally, socioeconomic position limits certain people's access to nutritious food, making diets "inequitable." Interventions to manage obesity must therefore take into account a number of environmental factors that have been linked to obesity.

For instance, several of the intervention programs in Latin America have included increased tariffs on goods high in sugar, marketing oversight, and consumer education to help them choose better food. The International Network for Food and Obesity Monitoring and Action Support (INFORMAS), a global network, implemented nine different actions to reduce obesity in Mexico, including providing and promoting healthy food, expanding the selection of healthy food, creating a comprehensive plan, and fighting obesity with funds raised from taxes on sugar-sweetened beverages, among other things.

Since the environment contains so many diverse edges, as was already mentioned, it cannot be simply studied. For instance, trying to sequence the entire environment using emerging technology like sequencing may take decades. Therefore, sequencing the human genome has been more straightforward than sequencing the environment. Understanding how multiple genetic variants contribute to illness susceptibility (polygenetic) or how genetic mutations can cause disease (monogenetic) will be much easier with the help of human genome sequencing.

CHAPTER TWO: HEALTH EFFECTS OF OBESITY

Diseases Caused by Obesity

People who are overweight or obese are more likely to develop significant diseases and health issues than those who are at a healthy weight. These include osteoarthritis, which causes the bone and cartilage of a joint to break down.

issues with breathing and sleep apnea.

several different types of cancer, including breast and colorectal.

a life of poor quality.

clinical depression, anxiety, and other mental diseases are examples of mental illness.

Pain in the body and trouble moving around.

Type 2 Diabetes

heart illness.

Asthma.

 fat in the liver.

elevated blood pressure.

To prevent these disorders, it's crucial to maintain a healthy weight with a balanced diet and frequent exercise. If you have particular worries, it is advised that you speak with a medical expert.

Emotional stability and psychological stability

A variety of practical and sociocultural variables might lead to mental health issues for people who are obese. These consist of:

Quality of time: Men and women who are very overweight frequently experience issues with physical and vocational functioning as a result of their size and recurring illnesses. Being physically unable to engage in their favorite activities, such as going to exciting events, traveling, or spending time with friends and family, can cause social isolation, loneliness, and a greater inability to deal with life's challenges. Depression has been linked to chronic pain by itself.

Weight bias and discrimination: Society's negative views on obesity are one of the largest obstacles for people who struggle with weight concerns. Weight bias refers to the views and stereotypes that characterize obese people as ugly, sluggish, and unmotivated. These adverse misconceptions may be pervasive in families, among peers, in the workplace, and among healthcare professionals in medical settings. They may result in discriminatory actions that have an impact on a person's self-worth, employment prospects, and even the caliber of their medical care.

Weight bias and low body image:
frequently go hand in hand. Patients might incorporate the societal stigma linked with obesity, making them feel self-conscious about their body weight and dissatisfied with their looks. People who suffer from being overweight may often worry about being scrutinized for their appearance.

Physiological problems: Obesity-related health issues can also adversely affect mental health. According to research, having too much body fat and eating poorly raises inflammatory indicators. This increased inflammation also affects the health of the immune system and raises the risk of depression.

The signs of obesity

Common Obesity Symptoms in Adults

Adult obesity symptoms frequently include:
extra body fat, especially at the waist
breathing difficulty
more perspiration than usual
Snoring
difficulty sleeping
Skin problems resulting from moisture accumulation in skin folds
Inability to carry out simple physical activities that you could do without difficulty before weight gain.
varying degrees of mild to severe fatigue.
especially in the joints and back
mental health problems include low self-esteem, depression, shame, and social isolation.

Common signs of childhood obesity include:

Fatty tissue deposits, which may be visible around the breast.
(Dark velvety skin around the neck and other places)
Stretch marks on the hips and back
Having trouble breathing when exercising

Slumber apnea
Constipation
Disease of the gastroesophageal reflux.

A low sense of self.
Boys facing delayed puberty, and girls encountering early puberty issues with the joints, such as flat feet or dislocated hips
The prevalence of childhood obesity varies by demographic group. For instance, children from lower-income homes are more likely than those from higher-income families to be obese.

CHAPTER THREE: THE HORMONE-RELATED FACTORS IN OBESITY

Insulin resistance's consequences

Insulin resistance, also known as a reduction in insulin sensitivity, happens when cells in your muscles, fat, and liver do not respond to insulin properly. Insulin, a hormone manufactured by your pancreas, is essential for sustaining life and controlling blood glucose (sugar) levels.

Insulin resistance can either be acute or chronic, and it occasionally can be treated. Insulin resistance plays a crucial role in the hormonal mechanism causing obesity. When cells develop insulin resistance, the body responds by producing more insulin. The overproduction of insulin promotes obesity and has several negative effects, including excessive fat accumulation (glucose is stored as fat), increased hunger, cravings for sweet and high-calorie foods, hormonal imbalance, and long-term, low-grade inflammation linked to obesity and other health problems.

The two main causes of insulin resistance are too much body fat, particularly in the abdominal area, and inactivity.

The impact of estrogen on body fat storage

Putting on weight can be difficult, especially when internal factors in our bodies—like hormones—are at play. Estrogen is one of those hormones, and when our estrogen levels are out of balance, it can start to change several variables that affect our weight. Either too little or too much estrogen in the body can cause weight gain or fluctuation. This is so because estrogen affects insulin sensitivity, glucose metabolism, and lipid metabolism.

Each of these variables has the potential to affect body weight by affecting things like appetite, hunger, satiety, and energy levels. Either too little or too much estrogen in the body can cause weight gain or fluctuation. This is so because estrogen affects insulin sensitivity, glucose metabolism, and lipid metabolism.

Each of these variables has the potential to affect body weight by affecting things like appetite, hunger, satiety, and energy levels.

Estrogen is a key player in several critical body processes in addition to promoting healthy sexual development and reproduction. This involves the metabolism of bones, cholesterol, glucose, insulin sensitivity, and fat storage in the body. It is also recognized that estrogen aids in controlling mood and brain activity. It only seems reasonable that our bodies (and weight) can be sensitive to variations in our estrogen levels given the important function estrogen plays in so many facets of our health.

For instance, estrogen dominance, also known as excessive estrogen levels in progesterone, might cause the body to create more insulin. In turn, this causes weight gain, elevated blood sugar, and insulin resistance. On the other hand, if estrogen levels are too low for other hormones, the body may start storing more energy as fat to maintain normal estrogen levels.

Mood swings, insomnia, anxiety, melancholy, and even infertility are all common signs of an estrogen imbalance, and they can all have an indirect effect on things like how hungry we feel, how much energy we have, and how successful we are at keeping a healthy weight.

Body structure and testosterone

When it comes to losing excess weight, testosterone is frequently one of the factors that is most disregarded. It's crucial to realize that it has a very close and fundamental connection to body weight.

Aside from masculine traits like facial hair and a low-pitched voice, testosterone also helps men develop stronger muscles and more dense bones, both of which are essential for the production of testosterone.

A healthy testosterone profile will keep you active by promoting body fat loss, giving you the ideal physique you have been striving for.

Having said that, if you intend to engage in a rigorous weight-loss regimen, attempt to keep your testosterone levels stable. One of your main strategies for getting the desirable waistline should be this.

A versatile hormone, testosterone regulates several body functions simultaneously. It has a significant impact on how well the male body functions generally, including how fat is distributed, how muscles and bones are built, and even how men's mental health is affected.

Testosterone and Fat Storage

The method by which fat will be stored in the body is decided by testosterone. more specifically, the body's stored fat. Speaking about fat distribution, you should be aware that a man's body has different areas where fat accumulates than a woman's body does due to differences in testosterone levels. For instance, men who have lower amounts of testosterone tend to gain weight by having more belly and abdominal fat.

In addition, it may be challenging to lose that fat quickly, which could eventually result in health issues brought on by being overweight. Aromatase is a secretion that the body's fat cells make that interacts with testosterone to cause hormonal imbalance. It is well-recognized that hormonal imbalances contribute significantly to weight gain and difficult-to-lose fat.

Muscle Gain

The hormone testosterone is one of the primary factors in muscular strengthening and bulking up. Your body's ability to maintain its weight is significantly influenced by your muscle mass. This means that the more muscle mass you have, the more calories you will burn off because muscles require a lot of calories to maintain themselves.

Testosterone is released more often when you exercise. The growth hormone is subsequently activated and increased cell development occurs as a result of the hormone's interaction with the muscles, causing the muscles to contract. Thus, higher testosterone levels also aid in the regeneration of damaged tissues, hastening the recovery of any injury.

Mental Health Impact

Testosterone has a strong correlation with brain function, which has an impact on mental health, according to several studies and research. The brain's behavior may be negatively impacted by a decrease in this hormone level. This can impair the emotional health of the brain, leading to symptoms such as poor mood, depression, weariness, lack of motivation, and anxiety.

Low Testosterone Production Causes

Your testosterone levels will change throughout your life, just like any other hormone. Age, heredity, lifestyle, the place you live in, and other social influences are just a few of the variables at play. Because of these, a person may experience a rise in the hormone's production at one stage of life and a sharp decline at another.

If you're lucky, there might be some treatments that can restore your body's natural testosterone production, but this again depends on the underlying cause. In other circumstances, it might not be curable, in which case you might have to accept it as is. Having said that, let's enquire a little more deeply into the many underlying causes of a decrease in testosterone production:

Aging

Aging is one of the most typical causes of decreased testosterone. When a guy hits puberty in adolescence, through the teenage years, this hormone is often at its highest level.

There may be a steady decrease in output as you age, especially after the age of 30. The synthesis of testosterone can be negatively impacted by other health and medical issues that are frequently associated with aging, such as weight gain, diabetes, etc. That brings up the following element.

Health Issues

Normal hormone function is maintained by excellent health. Sadly, we have no control over some illnesses. Men who have testicular disorders, nervous system or pituitary disorders, physical injuries, diabetes, cancer, or even some illnesses see a decrease in testosterone production.

Genetic Disorders

Genetic disorders like Klinefelter Syndrome frequently cause insufficient testosterone production. The malformation is caused by a genetic defect rather than something that is inherited. Depending on the specific circumstance, a testosterone increase or replacement may occasionally help the issue.

Drugs And Steroids

Some drugs, particularly steroids used to treat other medical disorders, might negatively impact testosterone levels. It can happen in any stage and at any age. While when the medicine is stopped in some circumstances, you might start producing normally again, in others, it might be a chronic problem.

Improper sleep

Another important aspect affecting testosterone production is sleep quality. The general quality of sleep is important for both physical and mental health. Numerous health issues, including a decrease in testosterone production, can result from persistent sleep abnormalities or insufficient sleep.

Low Testosterone Side Effects

weight gain in the body.

While there may be several causes for a rapid increase in weight, low levels of testosterone, which aid in metabolism, may be one of the main ones.

Check your hormone levels if you discover that you are consistently gaining weight despite following diet plans and exercising.

Difficulty In Focussing

If you find it restless and hard to concentrate on your work or often become forgetful, then low secretion of testosterone can be the reason behind it. A Correct diagnosis can uncover the real cause.

Lack Of Energy

Well, feeling exhausted after a long day's work can be normal. However, if you are tired all the time despite taking ample amounts of rest, then a decline in testosterone production can be the reason. Metabolism is regulated by testosterone leading to the release of energy from consumed food.

Anxiety and depression

Changes in mood are yet another sign of having a low level of testosterone. Numerous physical issues, including weight gain, poor sperm production, erectile dysfunction, etc., can have an impact on mental health even if they may not immediately affect mood. As a result, low testosterone levels are frequently linked to anxiety and depression.

Methods For Increasing Testosterone

Some effective methods for boosting your body's testosterone production include:

A balanced diet

The first thing you should change is your eating plan if you want to raise the amount of testosterone in your body. Reduce your consumption of all foods high in refined sugar and saturated fats, such as fruits, vegetables, leafy greens, fish, etc. In order to manage testosterone, you must consume foods high in vitamins B6, C, potassium, magnesium, K, D, and bromelain.

Exercises for Powerlifters

Weight training will increase your tolerance level, develop your muscles, and help you gain muscular mass. As muscle mass increases, testosterone is released proportionately, therefore as muscle mass increases, so does the hormone. After or after an exercise session, testosterone tends to release more.

Testosterone Supplements

Although prioritizing natural methods such as a healthy eating and consistent physical activity is optimal for addressing low testosterone issues, they might prove insufficient. This relies on the personal physique and the cause behind reduced hormone release. Taking reliable testosterone supplements that are certified, organic, and made with natural herbs will likely tackle the problem.

Improving Sleep Quality

Many people sacrifice their sleep in today's busy lifestyle. Poor quality sleep causes stress which can not only affect the production of testosterone but also the overall health of a person. Therefore, proper sleep management and improving the quality of sleep is one of the ways you can reduce stress and thus regulate your testosterone.

A healthy level of testosterone is a sign of excellent health. Testosterone directly contributes to weight gain, and when this hormone is released less, weight gain results. This hormone may be regulated and you can stay fit by doing regular exercises, building and strengthening your muscles, and eating a good diet.

CHAPTER FOUR: APPROACHES TO PREVENT OBESITY

Food consumption and nutrition options

Consider whole foods including fruits, vegetables, whole grains, lean meats, and healthy fats when choosing food for a balanced diet. The consumption of sweetened beverages and snacks that include a lot of added sugars causes weight gain. To reduce saturated and trans fats present in fried foods and processed snacks, one should consume healthy fats instead, such as those found in avocados, almonds, and olive oil. Fish, beans, poultry, and tofu are examples of lean proteins that help with satiety and muscle maintenance.

Vegetables, fruits, legumes, and whole grains are examples of foods high in fiber that help with digestion and promote feeling full. To promote your general health and control hunger, pay attention to your eating habits, enjoy your meals, and remain hydrated by drinking lots of water throughout the day. Vegetables, fruits, legumes, and whole grains are examples of foods high in fiber that help with digestion and promote feeling full. To promote your general health and control hunger, pay attention to your eating habits, enjoy your meals, and remain hydrated by drinking lots of water throughout the day.

Medical procedures

If your own efforts to reduce weight have been unsuccessful or you have other medical concerns that necessitate lowering weight, you may need to contact your doctor for assistance. If you have health issues due to obesity, you might need to take prescription medication. In an effort to lose weight, many people take over-the-counter or complementary medicine pills. However, you need to change your negative habits and eating patterns if you want to reduce weight permanently. The goal of behavior modification is to eliminate unhealthy eating habits.

Additionally, it boosts the daily amount of activity you get. Obesity-related eating disorders necessitate therapy and perhaps medication. Your healthcare provider will decide the best course of action for you after considering your age, general health, physical capabilities, and medical history. Past weight loss attempts, your level of obesity and your response to various treatments, procedures, and medications, your predictions for how the condition will progress, and your preference or opinion.

Obesity medical therapy options

Medications on prescription

The most often recommended medications function by either preventing the absorption of fat or by inducing a sense of fullness. Orlistat is one of the medications that is most frequently administered. About 30% of the fat you consume is blocked by orlistat as it passes through your digestive tract. With this medication, frequent, oily bowel motions are possible. However, symptoms frequently improve if you reduce the quantity of fat you consume.

If you don't make other dietary and exercise adjustments after stopping this medication, you can gain back some or all of the weight you lost. Orlistat frequently causes unpleasant side effects and may not be effective for everyone. After the FDA determined that another medication, sibutramine, may increase the risk of heart attack, the manufacturer pulled the drug off the market in October 2010. The drug lorcaserin is another one. It increases serotonin levels in the brain to decrease appetite. Phentermine and topiramate, a combo drug, also suppress appetite. Other prescription drugs are rarely utilized for more than a brief period of time. The potential for addiction and substance abuse makes amphetamines unadvisable. Always with your doctor before beginning any weight-loss medication.

Supplements

Many over-the-counter supplements claim to speed up fat-burning or decrease hunger. Some supplements can have harmful side effects. Many of the side effects, advantages, and hazards of these products haven't been thoroughly investigated because many of them—possibly the majority—haven't undergone clinical studies.

If the maker's claim seems nearly unattainable, it probably is. (For example, "Burning fat during your sleep at night!"). A component found in ephedra (ma-huang) is used in asthma medications. The FDA has outlawed dietary supplements that contain ephedra due to the possibility of harmful side effects. Laxative products have the potential to lower the potassium level in your blood. Heart and/or muscular issues could result from this. Popular product pyruvate may cause a small amount of weight reduction. The effects of consuming extra pyruvate, which is present in red wine, cheese, and red apples, have not been well investigated. Its promise for weight loss hasn't been shown scientifically.

A daily multivitamin can assist in filling the nutritional gap even for those who maintain a balanced diet because no product can replace eating a healthy diet. However, vitamin pills won't aid in your weight loss.

Before taking these supplements, always see your doctor because they may have a lot of negative effects.

Changes in behavior

Most obese adults who lose weight over time risk putting it back on if they don't alter their approach to healthy eating. You have several options for altering your behavior. Maintaining a food journal is one method. You record what you ate, where you ate it, and when you became hungry for it. You can record when and how long you exercised in an activity log. These diaries might assist you in identifying your food and exercise routines so that you can decide what needs to change. Techniques for behavior modification may benefit from the assistance of a counselor or psychologist. You can alter your perspective on body image with the use of these approaches.

You can continue working toward your weight loss objectives by using a reward system without food. Other behavioral suggestions include never watching TV, reading, or engaging in another activity while eating, and serving food family style rather than directly from the stove. Additionally, you can limit your portion sizes, use smaller plates, drink water with each meal, and make precise plans for losing weight.

Inpatient counseling for eating disorders

A person is not classified as a compulsive overeater solely by consuming a large amount of food at once. Everybody occasionally indulges too much.

However, some overweight persons binge and purge. Purge refers to self-induced vomiting or using laxatives to eliminate excess calories from binge eating. Some persons indulge in excessive food intake without vomiting. These behaviors are eating disorders that necessitate medical assistance. Obese or overweight individuals typically have these problems. Before attempting to reduce weight, it's crucial to get treatment for any eating disorders you may have.

Physical therapy may be recommended as a treatment for the following eating disorders:

Bulimia: A disorder in which individuals eats compulsively and vomits using diuretics, or intense exercise to prevent weight gain. The binge is frequently followed by feelings of guilt, humiliation, and melancholy.

Binge eating syndrome: A condition comparable to bulimia. There are times when the person binges or overeats. It varies from bulimia in that those who have it don't vomit, use laxatives, or use diuretics to rid their bodies of the extra food.

Nighttime eating: Waking up at midnight to eat is potentially risky and also a sign of abnormal eating habits. A person suffering from an eating disorder can get assistance from medical professionals, therapists, and dietitians. Counseling can assist in altering attitudes and behavior. A few people benefit from medication and support networks.

Exercise regimens: People who are new to exercising should speak with a doctor first to ensure that their plan is healthy for them. Generally speaking, it is advised to:

Begin slowly: It was discovered that any form of exercise dramatically reduces the risk of cardiovascular disease. Therefore, even if someone cannot yet exercise for long periods, even 5 to 10 minutes can be beneficial.

Appropriate movements: Avoid attempting overly demanding activities. Instead, people can modify movements to suit their degree of fitness. For instance, it could be simpler to walk more slowly or on a flat terrain than it is to run or climb a hill. As a person gains fitness, these adaptations enable them to challenge themselves more.

Think about low-impact exercise: Exercises that are easier on the body, including yoga, water aerobics, and walking on soft surfaces, may be helpful for people with joint discomfort or other ailments.

Include movement in your daily activities: Walking up and down stairs, doing housework, gardening, playing with pets or kids, and other activities can all be considered kinds of exercise.

Consider enrolling in a course: Group exercise can be motivating, teach someone how to perform specific exercises properly, and provide access to a fitness instructor's knowledge.

Think about physical therapy: If someone has chronic discomfort or hasn't worked out in a while, they might profit from a physical therapist's individualized care. It can be beneficial for those who are new to fitness to try relatively easy forms of exercise, such as:

Rotations of the trunk: Sit or stand with the arms at your sides. Move only your upper body; swing your arms and trunk from side to side.

Sit-to-stand: Take a seat close to the edge of a sturdy chair without armrests. Stand up, take a breath, and then take another breath. Sit back down carefully after the subsequent exhalation. Hold your arms out to the sides while seated or standing and perform arm circles. Move the hands in circles, from large to little, while maintaining straight arms.

Arm raises: While seated or standing, slowly lift your arms in front of your body before lowering them. They can also be raised over the head or out to the sides.

Sit or stand with your arms at your sides: Rotate your trunk. Just your upper body should be moving; your arms and trunk should be swinging. Take a seat near the edge of a strong chair without armrests to perform a sit-to-stand. Take a breath, then stand up and take another breath. Then exhale slowly and deliberately and sit back down.

Standing or sitting, make arm circles while holding your arms out to the sides. While keeping the arms straight, do small to large circular motions with the hands.

Arm raises: Raise and drop your arms slowly in front of your body whether seated or standing. They may also be held out to the sides or elevated above the head.

Stepping or marching: Try a quick walk outside, a slow march inside, or stepping up and down on a low stool. These workouts can be customized by changing the repetitions, pace, or intensity to fit the individual.

Home workouts to combat obesity.
Many of the beginner workouts listed above can be done at home. People can try these for more intensity:

Scaling stairs: This has the potential to be both strengthening and aerobic, especially for the legs. Stairs can be climbed and descended by walking or running.

Weight-training exercises: These are exercises that strengthen bones and muscles without the use of weights by using a person's body weight. Examples include squats, lunges, and leg raises.

Swimming.

People with knee or back discomfort may also need to avoid particular activities, such as twisting or jarring motions, depending on the underlying cause. Before beginning an exercise program, those with any underlying diseases, chronic pain, or disability should consult their doctor.

A person should also consult a doctor if any of the following apply: exercise exacerbates any medical issues; exercise hurts; a person feels physically incapable of exercising; a person needs assistance achieving certain health or fitness goals.

No one activity for obesity will be effective for everybody. To design a workout program that achieves a person's goals, it is important to consider their fitness level, general health, and personal preferences. Exercise should be difficult but not harsh or agonizing. It may be entertaining or empowering. Numerous facets of health can gradually improve as a result of time. A person may want to consult a doctor or physical therapist if they are unsure of where to start or are experiencing trouble exercising. As an alternative, students can enroll in classes or utilize free internet tools to experiment with various forms of exercise.

CHAPTER FIVE: HOW EXCESSIVE SUGAR CONSUMPTION CAN LEAD TO OBESITY AND FAT AMOUNT, SPECIFICALLY IN THE ABDOMINAL AREA

sugar addiction and sugar cravings

Obesity is influenced by excessive sugar intake and desires in many different ways. When consumed by a person, sweet foods and beverages frequently have a high-calorie count and have little to no nutritional value. These are empty calories that can result in an excessive daily calorie intake and contribute to weight gain.

Consuming sugar can induce sharp blood sugar rises and subsequent decreases. These changes can increase hunger, increase the desire for sweet foods, increase overeating, and increase the consumption of empty calories, which can increase body fat and obesity. Sugar interferes with the normal regulation of fullness and appetite hormones, making it more difficult for people to control their eating patterns.

To keep a balanced diet, it's crucial to keep an eye on your sugar intake to lower your risk of developing health problems associated with it.

The word "sugar" isn't usually present on food labels. It occasionally goes under different names, such as these:
Agave nectar
Brown rice syrup
sugary sweetened corn syrup
Dextrose
Juice from vaporized cane
Glucose
Lactose
Milky Malt Syrup
molasses
Sucrose.

Avoid products that contain more than 4 grams of sugar in total or that feature any type of sugar among the first few ingredients.

The consequences of additional sugar on energy loss and weight growth.

Fatigue and lack of sustained energy, increased hunger, insulin resistance and fat storage, metabolic imbalances, empty calories, and blood sugar spikes are some of the impacts of added sugar that contribute to weight gain. Limit added sugar intake and follow a balanced diet full of whole, unprocessed foods to maintain an appropriate level of energy and a healthy weight.

The internal fat storage is brought on by fructose consumption

One of the main organs affected by the negative consequences of high fructose consumption is the liver. Excessive fructose can cause non-alcoholic fatty liver disease, which is characterized by fat accumulation mostly in the liver. Because fructose and glucose are metabolized differently, excess fructose can be turned into fat by the liver. This causes liver issues and, in more serious situations, can result in complications like insulin resistance and metabolic syndrome. It is advised to keep an eye on your fructose intake to keep your liver healthy.

CHAPTER SIX: OBESITY IN DIFFERENT AGE GROUPS

Pediatric and adolescent obesity, or obesity in the early stages of life

When a child weighs more than is considered healthy for their age and height, it can lead to pediatric (childhood) obesity, a complex condition. The medical definition encompasses having a Body Mass Index (BMI) equal to or above the 95th percentile on the specialized growth charts established by the Centers for Disease Control and Prevention (CDC).

A child's BMI is different from an adult's. Because children's body compositions change as they get older, BMI is age- and sex-specific for them. Additionally, they differ between youngsters assigned to be male or female at birth. By dividing your child's weight in kilograms by their height in square meters (kg/m2), you may get their BMI.

For instance, a 10-year-old child's BMI would be 23.6 kg/m2 if they weighed 102 pounds (46.2 kg) and stood 56 inches (1.4 m) tall. They are obese since their BMI for their age puts them in the 95th percentile.

Obesity in children has a higher likelihood of persisting into adulthood. Children who are obese are more likely to suffer from a variety of illnesses. These circumstances include:
Asthma.
Apneic sleep.
Type 2 diabetes
Hypertension is a medical term for high blood pressure.
elevated cholesterol.
heart illness.
Stroke.
diseases of the muscles and bones, such as osteoarthritis.
some malignancies, including breast and colon cancer.
obese liver.

Children who are obese are also more likely to experience the following:
Bullying.
isolated from society.
A low sense of self.

Depression.

Pediatric obesity describes children aged 2 to 11 who have too much body fat. By setting an example for children, you can stop childhood. If your child sees you, they will emulate what you do. Being physically active and eating well will increase the likelihood that someone else will improve their behaviors. Plan family outings that might involve riding, swimming, or walking. Reduce your child's sugar intake if they are older than two. Less than 10% of their daily calories should come from sugar. Encourage your children, who are between the ages of 6 and 12, to get enough rest. Every night, they require 9 to 12 hours of sleep. Because it causes your child to desire to eat more and be less active, a lack of sleep can contribute to obesity. Ensure that your youngster visits their doctor annually as well. This might aid them in living a healthy lifestyle.

Adolescent obesity is the term used to describe people who are overly overweight when they are teenagers. It's a serious health problem influenced by several things, including sedentary behavior, a bad diet, heredity, and societal pressures.

It must be treated holistically, with healthy eating practices, regular exercise, education, and assistance from family members and medical specialists. Ages 12 to 19 are affected by adolescent obesity. Obesity in children and adolescents must be addressed with a comprehensive, balanced strategy that includes good food practices and regular physical activity.

in class 3 is commonly referred to as "severe" obesity.

Adult obesity poses a variety of problems and consequences, including a higher risk of developing chronic diseases like heart disease, diabetes, and several malignancies. Additionally, it might result in problems with mental health, increased healthcare costs, diminished mobility, and a lower quality of life. It will incorporate physical exercise, medical check-ups, and a better-balanced diet to address these issues.

Age-related obesity in the elderly

Regardless of age, anyone can become obese. However, as we age, obesity's traits and its effects on people can occasionally differ from those of younger folks. Knowing this is crucial because it could influence if and how older persons with obesity should be treated.

The Obesity Remedy

Obesity is a well-known and scientifically demonstrated risk factor for a wide range of illnesses. In actuality, obesity hurts the majority of body systems and organs. Patients who are obese are more likely to develop diabetes, hypertension, high cholesterol, heart disease, and several malignancies. Physical impairment is a significant issue as we age because of the impact of weight on joints. However, researchers have identified a phenomenon known as "the obesity paradox."

Obesity and overweight are unquestionably linked to a shorter lifespan while people are younger, but it appears that this is not always the case as people get older. According to several studies, elderly people may weigh more protectively than the "ideal" weight. Patients who are elderly and suffering from certain ailments appear to live longer when they are overweight or obese.

In the scientific community, there is continuous discussion about whether this is a real occurrence and, if so, what might account for it. Some claim that the statistics are what they are simply because those who are "susceptible" to the negative consequences of fat may have already passed away from illnesses as they aged. People who are "resistant" to the harmful consequences of fat thus represent the senior population that is afflicted by obesity. Let's use the comparison between lung cancer and smoking to help you better comprehend this.

Smoking and lung cancer.
You could occasionally hear of grandfather, who smoked all his life, yet is still in excellent health. It doesn't negate the fact that smoking has an impact on people's health. Grandpa is now older and doing well while still smoking like a chimney; everyone else has passed away from cancer or other lung problems at a younger age. He may have some form of tolerance for the negative consequences of smoking. This could help to explain the "obesity paradox" and why some obese older adults appear to fare better than their normal-weight peers. Despite this, additional research is required before it can be claimed with certainty whether obesity and being overweight are protective among older people. However, resistance to starting weight-loss programs in the elderly is occasionally observed, and this resistance may be caused, at least in part, by these ambiguities.

Additionally, obesity has an impact on cognition, which includes our capacity for remembering, comprehension, problem-solving, and decision-making. These abilities are known to decline with age, and studies show that people who are obese see these declines more quickly. This consequence of obesity is more critical than ever when people get older since good cognition enables the elderly to live fuller, more independent lives.

A lower quality of life has been conclusively associated with obesity. With age, this becomes increasingly more important. Elderly people typically already have several problems that lower their quality of life. Obesity merely increases the load.

The selection of weight-loss drugs is more constrained in older persons. This reduces the number of weight-loss drugs that are readily available. The main obstacles to providing weight-loss drugs to senior patients are side effects, pre-existing medical disorders, and combinations with other medications. Additionally, elderly folks are increasingly thinking of having bariatric surgery. To make the best choice and a satisfying decision, the medical team must carefully examine the patient's current health issues, surgical risks, and advantages of the surgery. As adults get older, it's crucial to make sure that they not only live healthier lives but also ones that are meaningful, honorable, and as independent as possible. Although obesity affects people at all stages of life, we should recognize that as we become older, the effects become more pronounced and that there are major advantages to treating it later in life.

CHAPTER SEVEN: THE CALORIC DILEMMA

Basics of Caloric Intake

The amount of energy in the food and drinks you consume is measured in calories. You can make informed choices regarding your diet and activity by being aware of calories.

The human diet contains calories mostly from three sources. They are derived from the three macronutrients: protein, fat, and carbs. The primary source is carbohydrates. They have four calories each per gram.

Per gram, they offer four calories. At nine calories per gram, fat comes in second and provides more than twice as many calories. The third source is protein, which has four calories per gram. (In some nations, food energy is measured in kilojoules rather than calories. However, the formula you require is 1 calorie = 4.2 kilojoules.)

Your Physical State

The notion that exercising burns calories is one that most people are familiar with. Your body, however, also uses energy just to stay alive. Your "basal metabolic rate," or BMR, refers to how much energy your body expends while at rest. It accounts for two-thirds of the calories you burn each day.

This is a sizable portion considering how much energy is needed for metabolism. Cell development, blood circulation, and body temperature regulation are just a few examples of the processes that make up metabolism. For a normal diet, your brain consumes 450 calories each day, or roughly 20 percent of your total intake.

You can consider these freebies if you are watching your calorie intake. You can better manage your diet and exercise regimen if you are aware of how many calories your body uses merely to stay alive.

When You Should Eat Calories

For your body to function properly and for your metabolic processes to be successful, you must consume a particular number of calories each day. This amount equates to about 2500 calories per day for men. Every day, women require about 2000 calories. Age, sex, weight, and exercise level are a few variables that affect these daily needs.

Here are some additional suggestions for increasing calorie awareness:
Drink more water with your meals. Consider starting with soup or drinking more water. In either case, drinking more liquids can make you feel satiated.
Boost your consumption of fiber. Because fiber is difficult to digest, the gut only receives half of the calories it contains. Fiber will also prolong your feeling of fullness.
If it isn't already a part of your routine, add protein to your breakfast. Protein provides you with prolonged fullness. Lean body mass can be supported by protein, which will assist in raising your BMR.

Avoid empty calories wherever you can. These are typically found in solid fats and added sugar. These foods provide you with energy, yet they are deficient in crucial nutrients.
For information on calories and macronutrient content, read nutrition labels. Pay close attention to how many serving sizes are included in each

package. The nutritional data only applies to one serving.

Bodyweight and Calories:

Weight control becomes a simple math problem when you are aware of the calories you consume from food (calories in) and the calories you expend (calories out). It takes a lot of work, but it's all about finding the right balance between calories in and calories out.

Paying attention to your nutrition will help you regulate your calorie consumption. Additionally, you can alter your calorie expenditure by including a variety of physical activities in your daily routine.
You will probably lose weight if your daily calorie intake is lower than necessary. You should keep your present weight if you meet the criterion. And you'll probably put on weight if your caloric intake exceeds your minimum requirement.

Calories In - Calories Out 0 = Loss of weight
Calorie maintenance: Intake minus expenditure equals zero.
Calories In - Calories Out > 0 = Weight Gain

Recall your age, weight, sex, and degree of activity as the variables that will affect your daily calorie requirements. These have an impact on how many calories you "take in" and, as a result, how many calories you need to burn off to attain your target.
Only about 100–400 calories can be burned through exercise per 30 minutes of activity, or 200–800 calories per hour. While the daily calorie requirement for the average person is about 2,250.

The "calories in" side of the equation is probably the simpler to influence right away. 500 calories require a lot of effort to burn off through activity. However, it's really simple to stop eating junk food, stop consuming empty calories, and reduce your daily calorie intake by 500. Not to minimize the advantages of exercise for health. Everyone should engage in at least four days of moderate-intensity exercise per week. You will be able to reap the advantages of exercise thanks to this.

Start slowly cutting back on your caloric intake to see the biggest difference in your weight. Increase your workout schedule if you first struggle with calorie restriction. But to reduce your weight as quickly as possible, focus on both sides of the calorie balance equation. Combine reducing your food intake, choosing healthy foods, and upping your regular activity regimen. It might be challenging to create a healthy routine that includes exercise and balanced food. Being aware of calories can be very beneficial. Fortunately, there are lots of tools at your disposal to make this process simpler.

Find out first what your energy requirements are depending on your age, sex, weight, and degree of activity. This provides you with a strong foundation. To make informed selections based on the labels on the food packaging, gather as much information as you can. By planning your meals and snacks in this way, you can meet your daily calorie needs. You may then determine how many calories you should be ingesting daily relative to your minimum energy requirements based on your weight management goals.

Decide how many calories you can burn based on the exercises you like to do. This can help you determine how frequently and for how long you should work out to reach your weight-management objectives.

The first step to becoming healthier may be to better understand calories. Whatever your objectives for health and fitness may be, you are more equipped to make choices now.

The value of considering factors other than calories

In addition to calories, nutrient-dense foods provide a variety of micronutrients that are important for maintaining good health. Because they frequently keep us feeling satiated for longer, nutrient-dense foods are also beneficial while trying to reduce weight. They are a crucial factor to take into account in addition to calorie tracking because they can help with weight management by improving a feeling of fullness and contentment.

Does eating too many calories necessarily lead to weight gain?

Over time, weight gain may occur if you consume more calories than your body requires. Additionally, weight gain occurs when the body stores additional calories as fat.

It depends on one's metabolism and other factors how much and how quickly one gains weight.

Low-fat versus low-carbohydrate diet

Many people turn to low-fat diets to aid in weight loss and increase fat burning.

However, recent studies suggest that low-carb diets might be equally effective. Moreover, low-carb diets have been proven to reduce hunger, promote fat loss, and sustain stable blood sugar homeostasis. You might be curious to know which option is more effective for weight loss.

fundamentals of each diet

Although both low-carb and low-fat diets are intended to aid in weight loss, they differ in their organizational principles.

a diet low in carbs

There are different levels of carbohydrate restriction in low-carb diets. They consist of:

Very low-carbohydrate diets: 20–50 grams of carbs per day on a 2,000-calorie diet, or less than 10% of total daily calories.

Less than 26% of total daily calories, or less than 130 grams on a 2,000-calorie diet, are from carbohydrates.

Low-carb diets: 26-44% of the daily caloric intake
Be aware that very low carbohydrate diets are frequently ketogenic, which means they severely limit carbohydrate consumption to cause ketosis, a metabolic state when your body burns fat for energy instead of carbohydrates.
Low-carb diets typically forbid items like baked goods, candies, sweets, and beverages with added sugar. In some variations, beneficial carbohydrate sources such as grains, starchy vegetables, high-carb fruits, pasta, and legumes may also be restricted.

Protein and fat intake should be increased at the same time, from healthy foods like eggs, non-starchy vegetables, nuts, fish, high-fat dairy, meat, and natural oils.

fat-free diet
In low-fat diets, fat intake is limited to less than 30% of daily calories.
Foods high in fat, such as butter, avocados, nuts, seeds, and full-fat dairy products, are frequently restricted or even prohibited.

You should eat naturally low-fat foods instead, such as fruits, vegetables, whole grains, egg whites, lentils, and skinless poultry. Lean cuts of beef and pig as well as low-fat yogurt and skim milk are occasionally acceptable fat-reduced foods.

It's crucial to keep in mind that some goods with decreased fat, like yogurt, may include additional sugar or artificial sweeteners.

Numerous studies have evaluated the effects of low-carb and low-fat diets on weight loss as well as several other elements of health to determine which is better for your health.

Loss of weight.
The majority of studies point to low-carb diets as possibly being more successful than low-fat diets for rapid weight loss.

An older, 6-month research of 132 obese individuals found that those who followed a low-carb diet lost more weight than those who followed a low-fat, calorie-restricted diet by more than 3 times.

In short 12-week research, overweight adolescents who followed a low-carb diet dropped an average of 21.8 pounds (9.9 kg), compared to those on a low-fat diet who lost only 9 pounds (4.1 kg). Similar to this, a two-year trial assigned 322 obese participants to a low-fat, low-carb, or Mediterranean diet.

Low-carbohydrate participants dropped 10.4 pounds (4.7 kg), low-fat participants 6.4 pounds (2.9 kg), and Mediterranean participants 9.7 pounds (4.4 kg).

Nevertheless, an alternative study implies that low-carb and low-fat can have similar long-term effectiveness.

Participants lost noticeably more weight on a low-carb diet than on a low-fat diet, according to a study of 17 research. After a year, the low-carb diet was still more effective, but the gap between the two gradually shrank.

A 2-year study of 61 diabetes patients indicated that low-carb and low-fat diets both caused weight changes.

Finding a diet that you can stick to may be the most crucial component for successful weight control, according to a comprehensive meta-analysis of 48 research that found that both low-fat and low-carb diets led to identical weight reduction.

slim down

The majority of studies show that low-carb diets are more effective at reducing body fat.

It was found during a 16-week trial with modest characteristics that low-carb and low-calorie dieters decreased their total body fat mass and abdominal fat more than low-fat dieters.

Similar results were found in a 148-person research conducted over a year.

Furthermore, several additional research indicates that low-carb diets are more effective at reducing belly fat than low-fat diets.

Finding a diet that you can stick to may be the most crucial component for successful weight control, according to a comprehensive meta-analysis of 48 research that found that both low-fat and low-carb diets led to identical weight reduction.

Appetite and hunger

In comparison to low-fat diets, studies generally suggest that low-carb, high-protein diets reduce sensations of hunger and enhance mood, making them possibly simpler to follow over the long run.

For instance, a low-fat diet was linked to higher decreases in levels of peptide YY, a hormone that suppresses appetite and encourages satiety, than a low-carb diet in a study of 148 persons.

This might be because protein and fat are satiating. Both of these macronutrients help you feel full for a longer period by delaying the emptying of your stomach.

Numerous hormones that regulate hunger and appetite have been proven to be impacted by protein and fat as well.

In a short study, meals high in protein and fat resulted in higher levels of the hormones glucagon-like peptide-1 (GLP-1), which signals fullness, and ghrelin, which signals hunger, to a larger extent than meals high in carbohydrates.

levels of blood sugar

Low blood sugar levels might make you feel more peckish and have major side effects like trembling exhaustion, and unexpected weight fluctuations.

One approach to help regulate blood glucose levels is limiting carb intake.

In one study of 56 persons with type 2 diabetes, it was found that a low-carb diet was superior to a low-fat diet in terms of regulating blood sugar, promoting weight reduction, and lowering insulin requirements. Only the low-carb diet was found to lower levels of circulating insulin, which in turn raised insulin sensitivity in smaller research including 31 individuals.

The ability of your body to get sugar from the bloodstream into your cells and maintain better blood sugar control can both be improved by increased insulin sensitivity. However, despite a 3-month trial in 102 diabetics showing that a low-carb diet is superior to a low-fat diet for weight loss, there was no discernible difference in blood sugar levels. Therefore, further study is required to determine how low-carb and low-fat diets affect blood sugar levels.

Different health repercussions.
Diets low in fat and carbs may have varying effects on various facets of health. These consist of:

Cholesterol. Low-carb diets are more successful than low-fat diets at raising HDL (good) cholesterol levels and lowering triglyceride levels, according to a study of eight trials. Both diets had a minimal impact on LDL (bad) cholesterol.
the heart rate. Although research indicates that both diets can lower blood pressure levels in the short

term, there is conflicting evidence about their long-term benefits.

Triglycerides

According to several studies, a low-carb diet can lower triglycerides more than a low-fat diet.

Insulin. Studies on the impact of low-carb and low-fat diets on insulin levels have produced inconsistent findings. To decide whether one is more advantageous than the other, more research is required.

People who choose low-fat diets are considered to be following a popular approach to weight loss.

Low-carb diets, on the other hand, are associated with greater short-term weight loss, as well as increased fat loss, decreased hunger, and improved blood sugar regulation.

Studies demonstrate that low-carb diets can be equally effective for weight reduction as low-fat diets and may also provide several additional health benefits, however, more research is required to determine the long-term consequences of each diet.

Remember that maintaining a consistent eating pattern over time is one of the most important aspects of success with both weight reduction and general health, regardless of whether you go for a low-carb or low-fat diet.

How much fat should you incorporate in your diet?

The right amount of fat for a healthy diet varies depending on factors like sex, degree of exercise, age, and general health. Good advice is to strive for 20–35% of your daily caloric intake to come from fats like those found in nuts, seeds, avocados, and fatty seafood.

A dietitian should be consulted for customized advice as well.

Which macronutrients should you think about incorporating in your diet?

Macronutrients, usually referred to as macronutrients, are nutrients that the body consumes daily in relatively substantial amounts. Proteins, carbs, and fats are the three macronutrients.

Protein: Aids in tissue repair and rebuilding while carbohydrates provide your body with energy.

Satiety, or feeling full, hormone balance, and the body's ability to absorb certain vitamins, such as vitamins A, D, E, and K, all depend on fat.

Your body receives quick energy from carbs. Amino acids, which are found in protein, are crucial for the synthesis of muscle, skin, blood, and significant brain and nervous system structures. Additionally, fat is essential for the growth of the brain, insulation, energy reserves, cell function, and organ protection.

Carbohydrates

The body prefers to use carbohydrates as fuel. The body finds it simpler to turn carbohydrates into usable energy than it does to turn fat or protein into fuel. Carbohydrates are essential for the health of your muscles, brain, and cells. Carbohydrates are transformed into blood sugar when you eat them (for instance, in the form of an energy bar before jogging). These sugars (in the form of glucose) can be used right away as a source of energy or they can be stored in the cells of the body for later use.

There are two types of carbohydrates: complicated and simple.
Long strings of sugar units make up complex carbohydrates (polysaccharides and oligosaccharides), which take the body longer to digest and utilize.

Complex carbohydrates Provide a more consistent impact on blood glucose levels.

The body can easily break down simple carbohydrates (monosaccharides and disaccharides), which are composed of one or two sugar molecules. Simple carbohydrates cause a brief influence on blood sugar levels. When ingested alone, some simple carbs, such as juice or sugary candies, can cause a rapid rise in blood sugar and energy levels followed by a rapid decline.

Complex carbs, especially fiber, not only give the body energy but also support good cholesterol levels and digestion.

Starchy foods like potatoes, rice, and grain products (including bread, cereal, and pasta) are examples of foods that are high in carbs. Carbohydrates can be found in fruits, vegetables, dairy products, and other foods.

Advanced Carbs

the legumes peas, beans, and others

whole grains

Cereals and bread

Rice

Veggies high in starch

Pasta

Table sugar Simple Carbs
Maple, honey, and other syrups
Candy
soda, sweetened tea, and fruit juice
Milk
Protein

Amino acids: which are the building blocks for muscle and other vital organs like the brain, neurological system, blood, skin, and hair, are provided by protein to the body. Additionally, protein carries oxygen and other vital nutrients. The body can reverse-process protein (a process known as gluconeogenesis) to use as energy in the absence of glucose or carbs.

On its own, your body can produce 11 amino acids. There are nine "essential amino acids" that your body cannot produce on its own, thus you must get them through your food.
You can obtain these amino acids by consuming various protein sources.

Complete Proteins
Give your body the right proportions of all the amino acids it needs. Complete protein is most

frequently found in meat, poultry, seafood, eggs, milk, quinoa, and edamame.

Some of the necessary amino acids are provided by incomplete proteins, but not all of them. Many proteins derived from plants aren't complete. However, you can obtain all the amino acids your body requires when you eat them together as complementary proteins.

Examples of incomplete proteins include nuts, seeds, and (most) cereals. To acquire the necessary amino acids you require throughout the day, you can eat these meals singly or in combination.

Complete proteins.

Chicken and eggs

Pork and beef

Salmon

Soy

Quinoa

Incomplete proteins

Lentils, Nuts, and Beans

whole grains

Vegetables

The amount of protein needed each day varies. Our daily calorie intake from protein sources should range from 10% to 35%, according to the USDA.

Based on age, sex, and activity level, more precise protein recommendations are given. To meet certain fitness or wellness objectives, some people will increase their protein intake.

Fats

Despite attempts to reduce fat in the diet, dietary fat is crucial for the body. In times of hunger or calorie restriction, fat serves as a vital source of energy.

Additionally, it is essential for insulation, healthy cell functioning, and safeguarding our key organs.

However, excessive calorie intake in the form of trans and saturated fat has been connected to several illnesses, such as diabetes and heart disease.

Understanding that fat has twice as many calories per gram as protein or carbs is crucial when choosing frozen meals or planning meals. You can consume a variety of fats as part of your everyday diet.

Dietary fats may be saturated or unsaturated in particular ways:

Meat and dairy products are the main sources of saturated fats. These fats often solidify at room temperature and have a lengthy shelf life. However, saturated fat from meats should be avoided as opposed to dairy when it comes to cardiovascular risk. Dairy products with full fat either do not affect cardiovascular health or have a positive effect.

Monounsaturated and polyunsaturated fats are two further forms of unsaturated fats. Unsaturated fats are found in plant-based diets, fortified foods like eggs and dairy, fish, seaweed, and goods made from grass-fed animals. They provide the body with a range of health benefits. These fats have a shorter shelf life than saturated fats and are typically liquid even when chilled.

The risk of developing certain ailments, such as heart disease, stroke, and type 2 diabetes, can be reduced when saturated fats in a person's diet are substituted with poly or monounsaturated fats.
Unhealthy Fats
Butter
Lean meats
Cheese
dairy items with added fat
unrefined fats
nut and seed
Olive oil is a type of plant oil.
Fish that is high in fat, like salmon and tuna
Avocado.

Trans fat is a different type of fat that is gradually being removed from meals. Polyunsaturated fats are metabolized to make trans fats, which are shelf-stable. These hydrogenated fats are commonly found in processed meals like crackers, cookies, cakes, and other baked items. The majority of dietary recommendations recommend that between 20% and 35% of your daily calories come from fats. However, you should consume no more than 10% of your daily calories from saturated fats.

Every macronutrient should be a part of your everyday diet. Building each meal around a mixture of protein, carbs, and healthy fats will make this easier. But it can be challenging to discover the precise macro balance that works for you. There is potential for experimenting given the wide range of suggested ratios for each macronutrient.

The way each person's body responds to different ratios varies.

Every macronutrient plays a critical function in the body. While some popular diets drastically limit or even eliminate certain macronutrients, each is necessary for your body to function optimally. Unless your healthcare professional has told you otherwise, such as if you are managing a medical condition, you should take each of them in moderation.

You can start learning how to make healthy decisions within each group once you've worked out how to balance your macros. Make lean proteins, complex carbohydrates, and healthy fats your go-to foods if you want to achieve your fitness objectives and stay healthy.

It's crucial to remember, though, that severe macro counting may not be advised for those with a history of eating disorders. Before making any big changes to your diet, it is essential to speak with your healthcare physician or a certified dietitian. This type of eating strategy also makes it harder for a person to pay attention to their internal hunger cues.

CONCLUSION.

This book looks closely at the causes of obesity, taking into account factors including thoughts, genes, culture, and environment. Since its discovery, obesity has been a serious health issue that affects many facets of daily life.

Now that you've gained awareness, you should push for significant improvements, encourage others to feel compassion and address this issue carefully. To combat obesity and make our society more egalitarian and healthy for future generations, we all need to work together, especially healthcare professionals and leaders.